Our Solar System
Earth in Space

by Glen Phelan

Table of Contents

Develop Language 2

CHAPTER 1 How Does Earth Move? 4
           Your Turn: Predict 9
CHAPTER 2 Earth's Place in Space 10
           Your Turn: Classify 15
CHAPTER 3 Rocks in Space 16
           Your Turn: Summarize 19

Career Explorations 20
Use Language to Compare 21
Science Around You 22
Key Words 23
Index 24

DEVELOP LANGUAGE

Describe the sun.

The sun is _____ and _____.

The sun looks _____.

What other objects do you see?

I see _____ and _____.

How are these objects alike?

They are alike because _____.

How are these objects different?

2 Our Solar System: Earth in Space

Photos and drawings in this book may not be to scale.

Photograph of the sun, taken on September 14, 1999

Develop Language 3

CHAPTER

How Does Earth Move?

We live on a **planet** called Earth. Earth moves two ways.

To rotate means to spin around like a top. Earth **rotates**.

Earth also **revolves** around the sun.
To revolve means to go all the way around something.

Earth rotates and revolves at the same time.

At all times, half of Earth is facing toward the sun.
On this half of Earth, it is day.

It is night in Miami.

At all times, half of Earth is facing away from the sun. On this half of Earth, it is night.

A place goes from day to night as Earth rotates.

Chapter 1: How Does Earth Move? 7

Earth's Seasons and the Sun

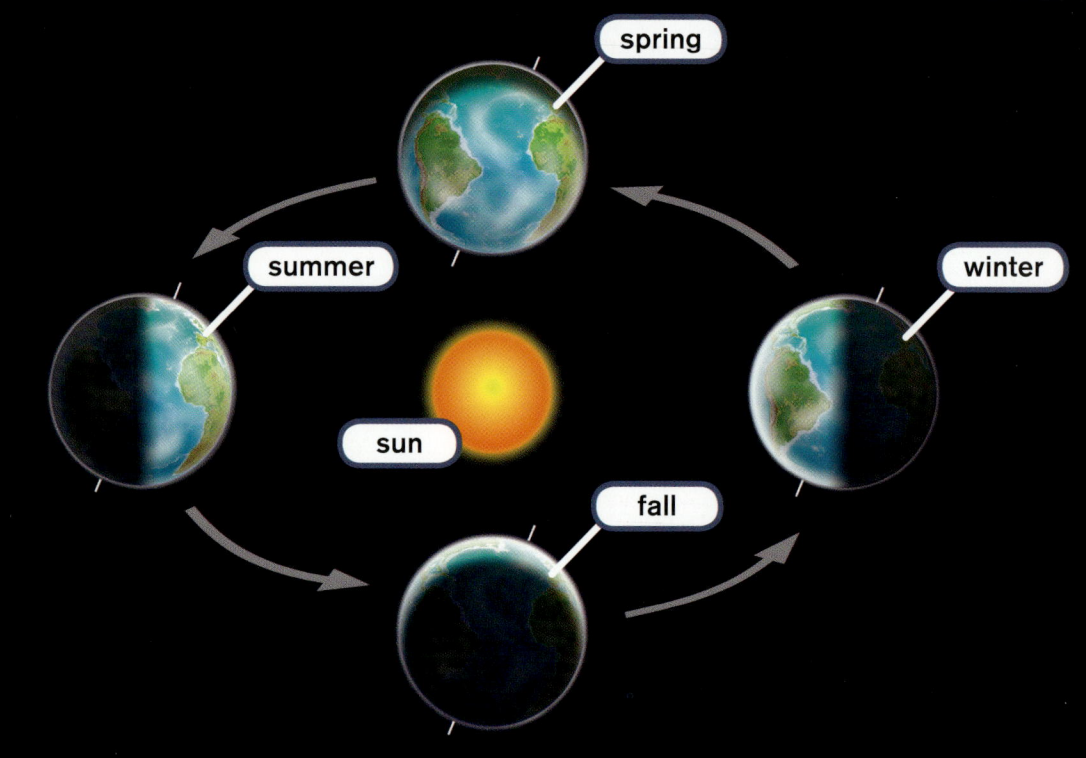

Earth is **tilted**. That means Earth is always leaning to one side.
It is summer in the part tilted toward the sun.
It is winter in the part tilted away from the sun.
The seasons change as Earth revolves.

KEY IDEAS Earth has day and night because Earth rotates. Earth has seasons because Earth is tilted as it revolves around the sun.

YOUR TURN

PREDICT

Look at the photo. What season will come next? Make a prediction.

I predict that _____ will come next.

Explain your prediction.

I predict this because _____.

MAKE CONNECTIONS

Where you live, how is summer different from winter?

USE THE LANGUAGE OF SCIENCE

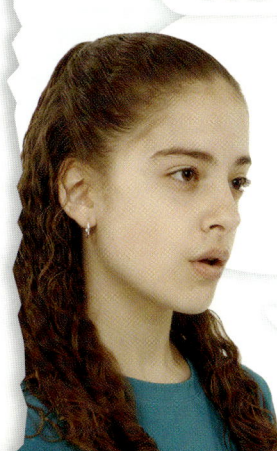

How does Earth move?

Earth rotates like a top and revolves around the sun.

Chapter 1: How Does Earth Move?   9

CHAPTER 2
Earth's Place in Space

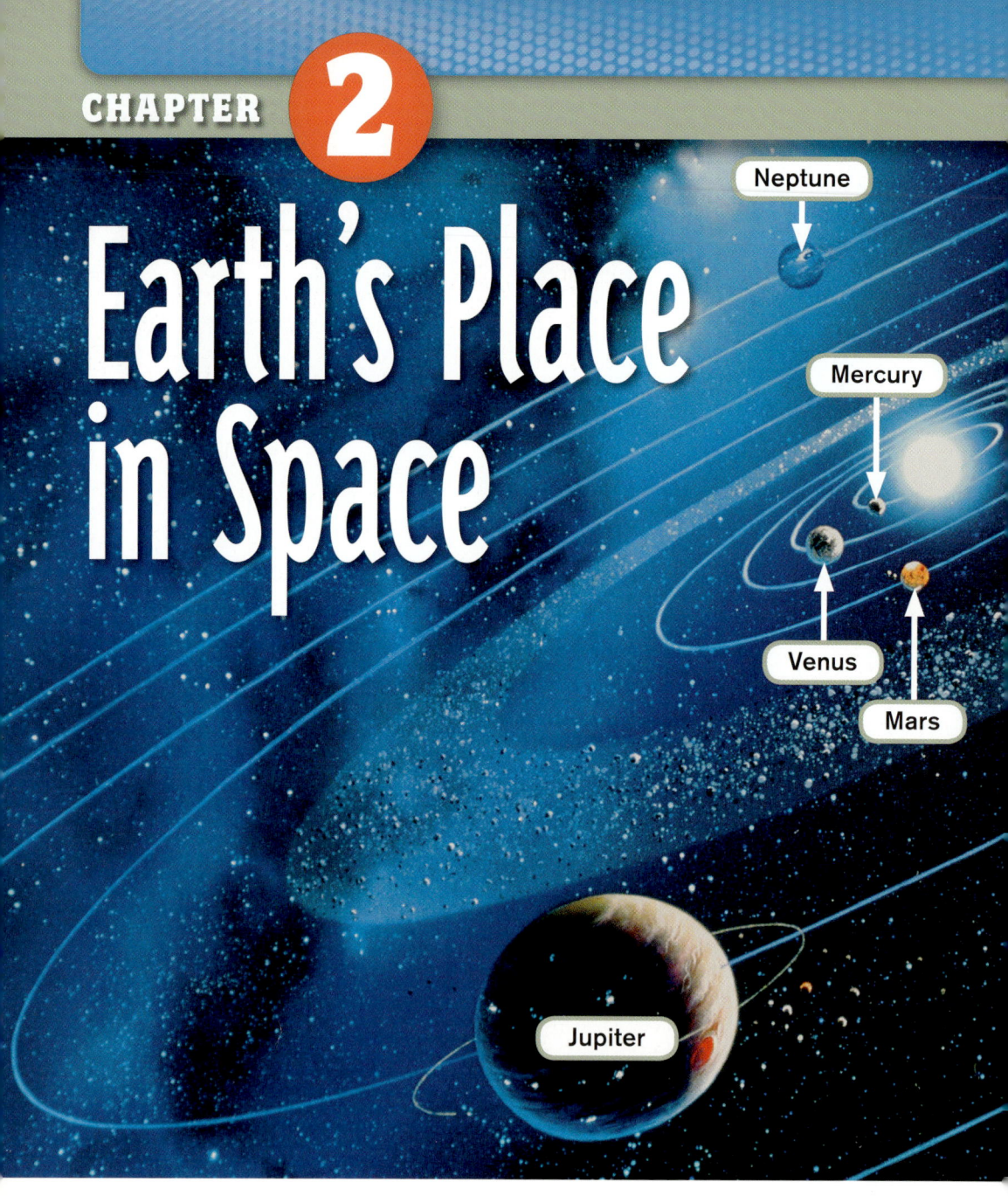

The sun is the center of our **solar system.**
Earth revolves around the sun.
Other objects revolve around the sun, too.

There are eight planets in our solar system. The planets revolve around the sun.

Chapter 2: Earth's Place in Space

Some planets in our solar system have **moons**.
A moon revolves around a planet.
Earth has one moon.

Earth's moon

12 *Our Solar System: Earth in Space*

The moon revolves around Earth.
The moon moves around Earth as Earth revolves around the sun.

Chapter 2: Earth's Place in Space 13

Some planets have no moons.
Other planets have many moons.
Jupiter has more than 60 moons!

▲ Ganymede is Jupiter's largest moon. It is the largest moon in our solar system.

KEY IDEA Planets and moons are part of our solar system.

14   *Our Solar System: Earth in Space*

YOUR TURN

CLASSIFY

Match these photos to other photos in the book. Classify each object by telling if it is a planet or a moon.

object A	moon
object B	
object C	
object D	

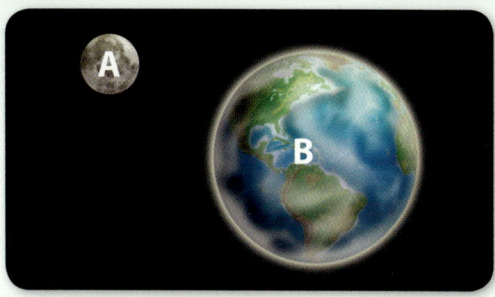

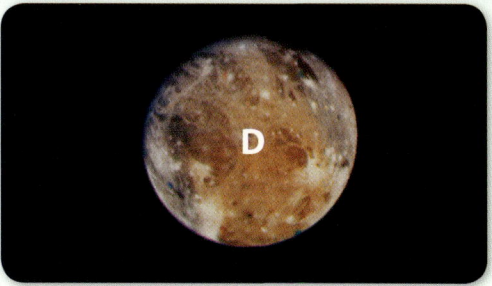

MAKE CONNECTIONS

When do you see Earth's moon?

STRATEGY FOCUS

Monitor Comprehension

What did you find hard to understand in this chapter? What did you do to understand it better?

Chapter 2: Earth's Place in Space 15

CHAPTER 3

Rocks in Space

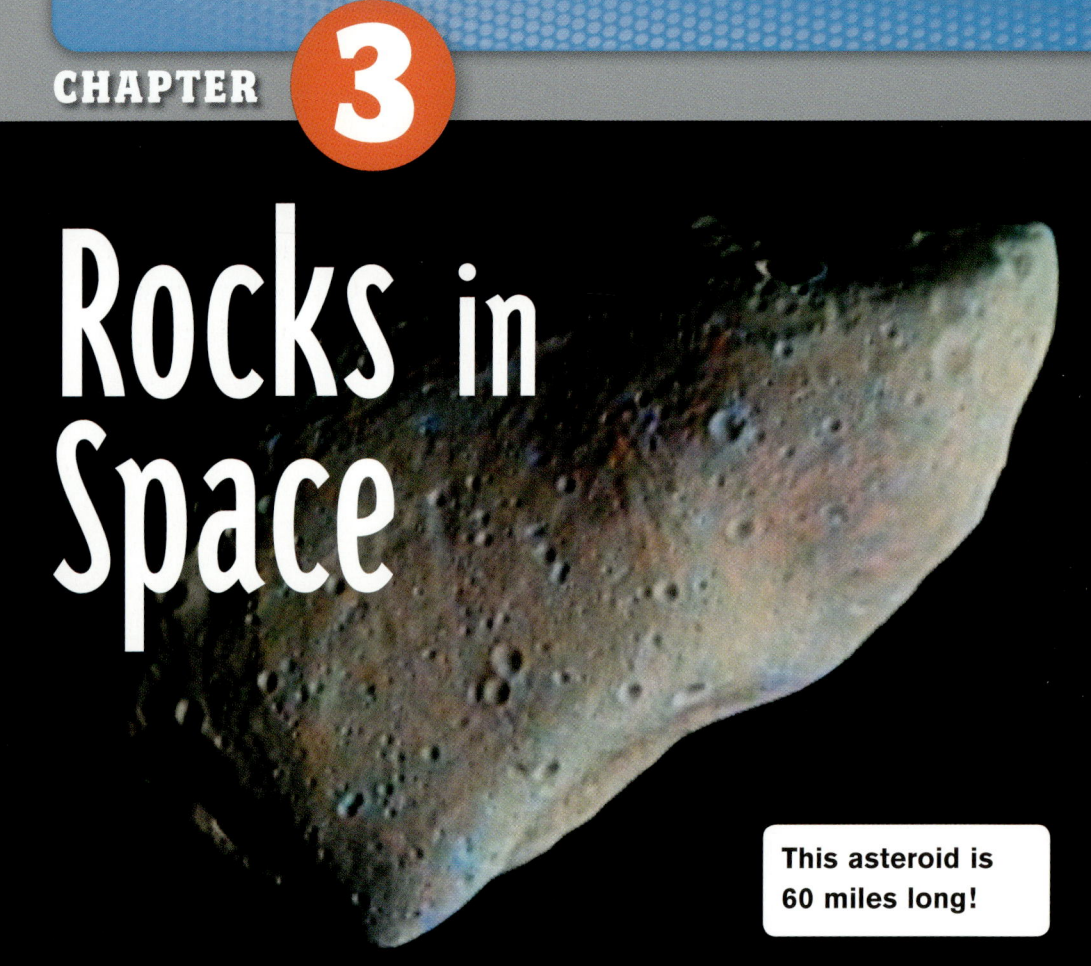

This asteroid is 60 miles long!

Planets and moons are part of our solar system. **Asteroids** are part of our solar system, too.

Asteroids are large chunks of rock in space. Millions of asteroids revolve around the sun.

Explore Language

GREEK WORD ROOTS

asteroid

aster (star) + *eidos* (form) = having the form of a star

falling rock

A small rock from space burns up as it falls toward Earth. It makes a bright streak in the sky.

Small chunks of rock also move through space. Sometimes these rocks reach Earth. The rocks often burn up as they fall toward Earth. They make bright streaks in the sky.

Chapter 3: Rocks in Space

Comets are part of our solar system.
A comet is a chunk of ice and rock.
Sometimes a comet gets close to the sun.
Then the ice on the comet turns to gas.
The gas forms the comet's tail.

> **KEY IDEAS** Asteroids and comets are rocky objects in our solar system. Everything that revolves around the sun is part of our solar system.

Your Turn

SUMMARIZE

This chapter is about rocks in space. Tell what you learned about some of these objects.

An asteroid is made of _____.

A comet is made of _____.

Asteroids and comets _____ around the sun.

Asteroids and comets are part of our _____.

MAKE CONNECTIONS

What objects have you seen in the night sky? Describe them to a friend.

EXPAND VOCABULARY

You have read about the solar system. The origin of the word **solar** is the Latin word *sol*. *Sol* means "sun." Find out what these terms mean:

    solar energy   solar car   solar panel

Find out what the word **lunar** means. Look up the origin of the word. Draw and write to share what you learn about these words.

Chapter 3: Rocks in Space

CAREER EXPLORATIONS

What Is an Astronomer?

▲ An astronomer uses telescopes and other tools.

▲ An astronomer learns about objects in space.

- Are you interested in learning more about our solar system?
- Would you like to be an astronomer?

An astronomer may work in an observatory. This is Yerkes Observatory in Wisconsin.

USE LANGUAGE TO COMPARE

Words that Compare

You can use **like** and **but** to compare things.
These words help show how things are alike and different.

EXAMPLE

Like Earth, the moon revolves.
But the moon revolves around Earth, not the sun.

You can use **too** to show how things are alike.

EXAMPLE

Planets and moons are part of our solar system.
Asteroids are part of our solar system, **too**.

With a friend, compare Earth and the moon.
Use **like**, **but**, and **too**.

Write a Comparison

You live on planet Earth. Find out about another planet in our solar system. Compare Earth and the other planet. Tell how they are alike and different.

Words You Can Use	
Comparison Words	
like	both
but	different because
too	alike because

Use Language to Compare **21**

SCIENCE AROUND YOU

FelipeNYC: Hi, Maria. It's winter here! It is really cold! Is it cold in Sao Paulo, too?

IMariaSP: Hi, Felipe. No. It's hot! When it's winter in New York City, it's summer here!

Read the e-mails from Felipe and Maria. Then answer the questions below.

1. Where do Felipe and Maria live?

Felipe lives in _____ , and Maria lives in _____ .

2. Why is it winter in New York City and summer in Sao Paulo?

_____ is tilted away from the sun.

_____ is tilted toward the sun.

Our Solar System: Earth in Space

Key Words

asteroid (asteroids) a chunk of rock that revolves around the sun
Some **asteroids** are many miles long.

comet (comets) a chunk of ice and rock that revolves around the sun
This **comet** has a long tail of gas.

planet (planets) a large, round object that revolves around the sun
Saturn is a **planet**.

revolve move around something
The moon **revolves** around Earth.

rotate spin around
A top is a toy that **rotates**.

solar system the sun and all objects that move around it
The sun is at the center of our **solar system**.

tilted leaning to one side
Earth is always **tilted** as it revolves around the sun.

Key Words 23

Index

asteroid 3, 16, 18–19
astronomer 20
comet 18–19
day 6–8
Earth 2–13, 15, 17, 21
gas 18
Jupiter 2, 10–11, 14
moon 12–15, 21
night 6–8

planet 2–4, 10–16, 21
revolve 5, 8–13, 16, 18–19, 21
rotate 4–9
Saturn 2–3, 10–11, 15
seasons 8–9, 22
solar system 10–21
sun 2–11, 16, 18–19, 21–22
tilted 8

MILLMARK EDUCATION CORPORATION
Ericka Markman, President and CEO; Karen Peratt, VP, Editorial Director; Rachel L. Moir, Director, Operations and Production; Mary Ann Mortellaro, Science Editor; Amy Sarver, Series Editor; Betsy Carpenter, Editor; Guadalupe Lopez, Writer; Kris Hanneman and Pictures Unlimited, Photo Research

PROGRAM AUTHORS
Mary Hawley; Program Author, Instructional Design
Kate Boehm Jerome; Program Author, Science

BOOK DESIGN Steve Curtis Design

CONTENT REVIEWER
Tom Nolan, Operations Engineer, NASA Jet Propulsion Laboratory, Pasadena, CA

PROGRAM ADVISORS
Scott K. Baker, PhD, Pacific Institutes for Research, Eugene, OR
Carla C. Johnson, EdD, University of Toledo, Toledo, OH
Donna Ogle, EdD, National-Louis University, Chicago, IL
Betty Ansin Smallwood, PhD, Center for Applied Linguistics, Washington, DC
Gail Thompson, PhD, Claremont Graduate University, Claremont, CA
Emma Violand-Sánchez, EdD, Arlington Public Schools, Arlington, VA (retired)

PHOTO CREDITS Cover © NASA Johnson Space Center (NASA-JSC); 1 © Digital Vision/Punchstock; 2-3, 3b, 16, 23a © NASA/JPL-Caltech; 2 © The Print Collector/Alamy; 3a and 23c © StockTrek/Getty Images; 3c © NASA/MODIS/USGS; 4a, 5, 6b, 7b 8, 21, 23g illustrations by Cam Wilson; 4b and 23e © D. Hurst/Alamy; 6a and 7a © Javier D. Fontanella/Shutterstock; 9a © Anyka/Shutterstock; 9b and 9c Lloyd Wolf for Millmark Education; 10-11, 23f © D. Van Ravenswaay/Photo Researchers, Inc.; 12, 13a, 23d © Brad Thompson/Shutterstock; 13b © Antonio Petrone/Shutterstock; 14 © NASA/Photo Researchers, Inc.; 15a © John R. Foster/Photo Researchers, Inc.; 15b © SPL/Photo Researchers, Inc.; 15c © NASA/Science Source; 17 and 24 © Frank Zullo/Photo Researchers, Inc.; 18 and 23b © Pekka Parviainen/Photo Researchers, Inc.; 19 © Derke/O'Hara/Getty Images; 20a and 20c © Roger Ressmeyer/Corbis; 20b © California Association for Research in Astronomy/Photo Researchers, Inc.; 22a © Joshua Haviv/Shutterstock; 22b © Celso Pupo/Shutterstock

Copyright © 2008 Millmark Education Corporation

All rights reserved. Reproduction of the whole or any part of the contents without written permission from the publisher is prohibited. Millmark Education and ConceptLinks are registered trademarks of Millmark Education Corporation.

Published by Millmark Education Corporation
7272 Wisconsin Avenue, Suite 300
Bethesda, MD 20814

ISBN-13: 978-1-4334-0098-8
ISBN-10: 1-4334-0098-7

Printed in the USA

10 9 8 7 6 5 4 3 2 1

Our Solar System: Earth in Space